金 星

炎热的剧毒星球

VENUS

The Hot and Toxic Planet

（英国）埃伦·劳伦斯／著 张 骁／译

江苏凤凰美术出版社

著作权合同登记图字：10-2022-144

图书在版编目（CIP）数据

金星：炎热的剧毒星球 /（英）埃伦·劳伦斯著；

张骁译 . -- 南京：江苏凤凰美术出版社，2025. 1.

（环游太空）. -- ISBN 978-7-5741-2027-3

Ⅰ . P185.2-49

中国国家版本馆 CIP 数据核字第 2024VH4391 号

策　　　划	朱　婧	
责 任 编 辑	高　静	奚　鑫
责 任 校 对	王　璇	
责任设计编辑	樊旭颖	
责 任 监 印	生　嫄	
英 文 朗 读	C.A.Scully	
项 目 协 助	邵楚楚	乔一文雯

丛 书 名	环游太空
书　　名	金星：炎热的剧毒星球
著　　者	（英国）埃伦·劳伦斯
译　　者	张 骁
出 版 发 行	江苏凤凰美术出版社（南京市湖南路 1 号 邮编：210009）
印　　刷	南京新世纪联盟印务有限公司
开　　本	710 mm×1000 mm　1/16
总 印 张	18
版　　次	2025 年 1 月第 1 版
印　　次	2025 年 1 月第 1 次印刷
标 准 书 号	ISBN 978-7-5741-2027-3
总 定 价	198.00 元（全 12 册）

版权所有　侵权必究

营销部电话：025-68155675　营销部地址：南京市湖南路 1 号

江苏凤凰美术出版社图书凡印装错误可向承印厂调换

目录 Contents

书中加粗的词语见词汇表解释。

Words shown in **bold** in the text are explained in the glossary.

欢迎来到金星
Welcome to Venus

想象一个距离地球数千万千米的酷热世界。

Imagine a scorching hot world that is millions of kilometers from Earth.

那里的地面遍布岩石，有超过1 000座火山。

The land is rocky, and there are more than 1,000 **volcanoes**.

虽然是白天，但是看起来却像黑夜。

It is daytime, but it looks like night.

这是因为有一层厚厚的云和有毒气体遮蔽了太阳光。

That's because a thick layer of clouds and poisonous **gases** blocks out the Sun's light.

欢迎来到金星！

Welcome to the **planet** Venus!

太空飞行器曾经造访过金星，它们将有关这颗行星的信息传回给地球上的科学家。这张图片是运用这些信息在电脑上生成的，它显示了金星表面的模样。

Spacecraft have visited Venus. They sent information about the planet back to scientists on Earth. This picture was created on a computer using that information. It shows how the surface of Venus might look.

人类是无法在金星上生存的。因为那里太热，而且无论对什么生物来说，那些有毒物质或毒气都是致命的。

A human could not survive on Venus. It is too hot, and the **toxic**, or poisonous, gases would be deadly to any living thing.

剧毒的云层和气体
Toxic clouds and gases

太阳系 The Solar System

金星以约126 000千米每小时的速度在太空中运动。

Venus is moving through space at about 126,000 kilometers per hour.

它沿着一条巨大的圆形轨道绕太阳运动。

It is moving in a big circle around the Sun.

金星是围绕太阳公转的八大行星之一。

Venus is one of eight planets circling the Sun.

八大行星分别是水星、金星、我们的母星地球、火星、木星、土星、天王星和海王星。

The planets are called Mercury, Venus, our home planet Earth, Mars, Jupiter, Saturn, Uranus, and Neptune.

冰冻的彗星和被称为"小行星"的大型岩石也围绕着太阳公转。

Icy **comets** and large rocks, called **asteroids**, are also moving around the Sun.

太阳、行星和其他天体共同组成了"太阳系"。

Together, the Sun, the planets, and other space objects are called the **solar system**.

太阳系中的大多数小行星都集中在被称为"小行星带"的环状带中。

Most of the asteroids in the solar system are in a ring called the asteroid belt.

太阳系 The Solar System

金星是距离太阳第二近的行星。
Venus is the second planet from the Sun.

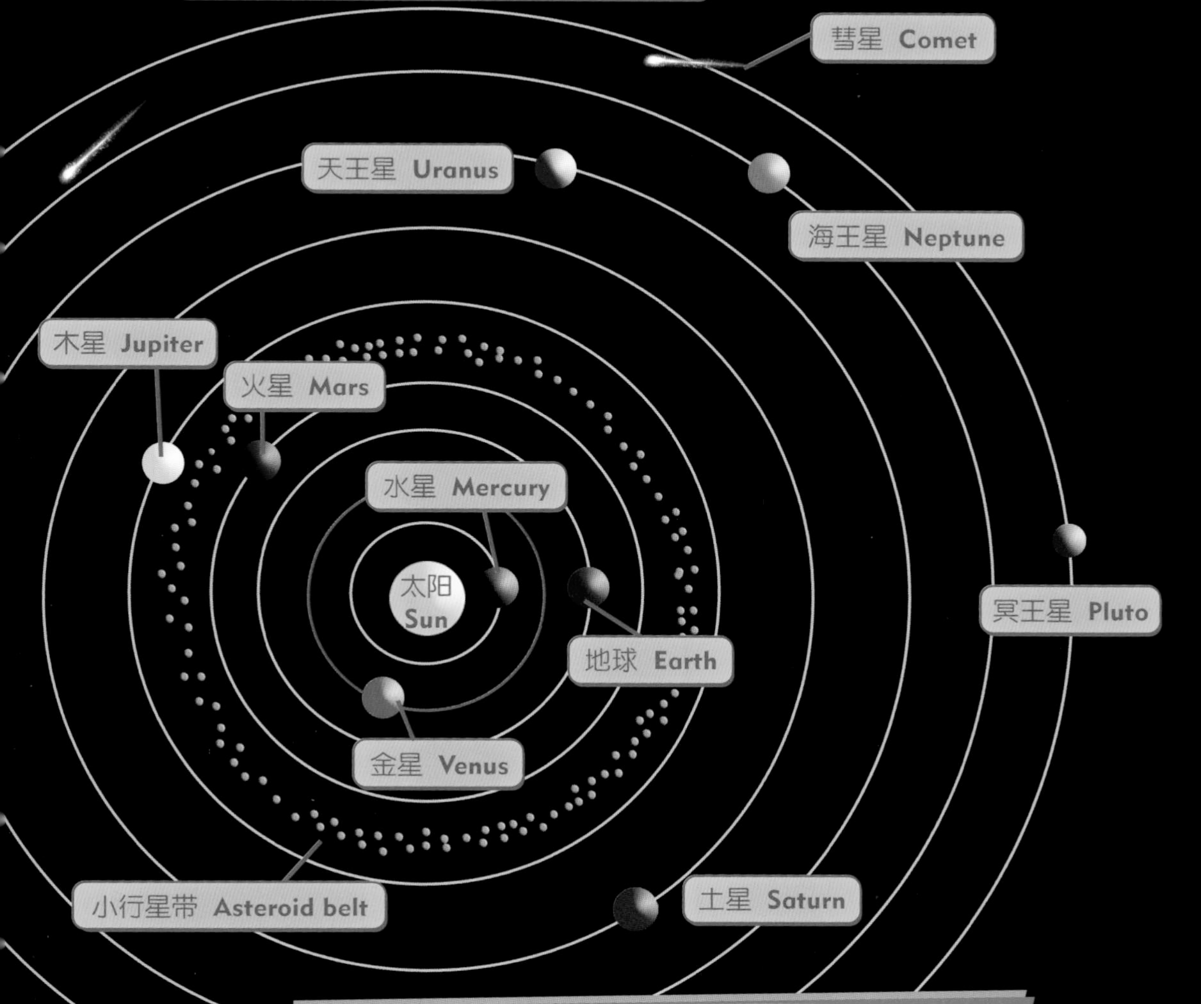

彗星 Comet

天王星 Uranus

海王星 Neptune

木星 Jupiter

火星 Mars

水星 Mercury

太阳 Sun

地球 Earth

冥王星 Pluto

金星 Venus

小行星带 Asteroid belt

土星 Saturn

太阳系里还有更小的星球，它们被称为"矮行星"。冥王星就是一颗矮行星。

The solar system is home to small planets, called **dwarf planets**. Pluto is a dwarf planet.

7

金星的奇幻之旅
Venus's Amazing Journey

行星围绕太阳公转一圈所需的时间被称为"一年"。

The time it takes a planet to **orbit**, or circle, the Sun once is called its year.

地球绕太阳公转一圈需要略多于365天，所以地球上的一年有365天。

Earth takes just over 365 days to orbit the Sun, so a year on Earth lasts 365 days.

金星比地球离太阳更近，所以它绕太阳公转的路程更短。

Venus is closer to the Sun than Earth, so it makes a shorter journey.

金星绕太阳公转一圈大约需要225个地球天。

It takes Venus about 225 Earth days to orbit the Sun.

这意味着，一个在地球上7岁的孩子，在金星上就有11岁了！

This means that a seven-year-old on Earth would be 11 in Venus years!

当行星围绕太阳公转时，它也像陀螺一样自转着。地球自转一圈要花24小时。金星自转得非常慢，转一圈要243个地球天。

As a planet orbits the Sun, it also spins, or **rotates**, like a top. Earth takes 24 hours to rotate once. Venus spins very slowly and takes 243 Earth days to rotate once.

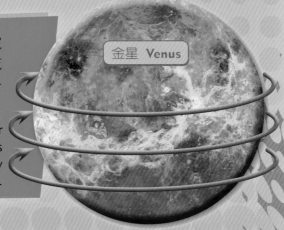

金星 Venus

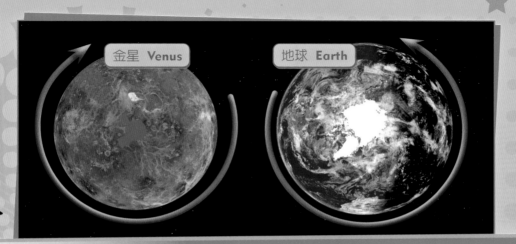

如果你从金星的上空俯视这颗行星，你就会看到它呈顺时针方向自转。而大多数行星都是以逆时针方向自转的。

If you could look at Venus from above the planet, you would see that it rotates in a clockwise direction. Most planets rotate in a counterclockwise direction.

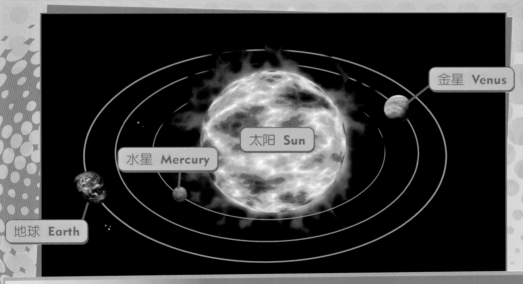

地球绕太阳公转一圈的路程约为9.4亿千米，而金星绕太阳公转一圈的距离大约为6.8亿千米。

To orbit the Sun once, Earth makes a journey of about 940 million kilometers. Venus makes a journey of about 680 million kilometers.

近距离观察金星
A Closer Look at Venus

金星是离地球最近的行星。

观察金星甚至都不需要用望远镜——你可以直接用肉眼看到它！

从地球上看，金星是太阳系中最耀眼的行星。

因为覆盖着一层厚厚的白金色云层，它看上去格外闪耀。

太阳的光芒在金星的云层上反射回照，使这颗行星闪烁发光。

Venus is the closest planet to Earth.

You don't need a telescope to see Venus—you can see it just with your eyes!

From Earth, Venus is the brightest planet in the solar system.

It looks so bright because of its covering of thick, yellowish-white clouds.

The Sun's light reflects, or bounces, off the clouds, making the planet shine.

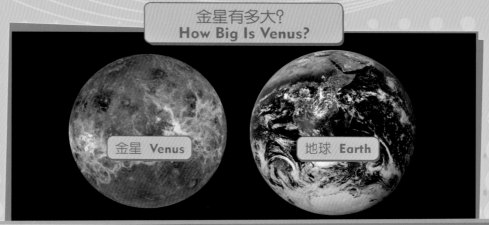

金星有多大?
How Big Is Venus?

金星 Venus

地球 Earth

金星只比我们的家园——地球稍小一点。这张图片显示了覆盖着云层的金星的样子。

Venus is just slightly smaller than our home planet Earth. This picture shows how Venus looks beneath its clouds.

月亮 The Moon

金星 Venus

这张金星的照片是在地球上拍摄的。在快要日出之前和刚刚日落之后，我们常常可以看到金星在天空中闪烁。

This photo of Venus was taken from Earth. Venus can often be seen shining in the sky just before sunrise or just after sunset.

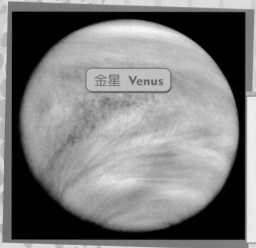

金星 Venus

这张有着厚厚云层的金星的照片是由一个太空飞行器拍摄的。云的顶层被像飓风般强大的气流吹散在金星周围！

This photo of Venus's thick clouds was taken by a spacecraft. The top layer of clouds is blown around the planet by winds that are as powerful as a hurricane!

致命的金星
Deadly Venus

我们的母星——地球，被一层厚厚的气体包围，这一气体层被称为"大气层"。

Our home planet, Earth, is covered with a thick layer of gases called an **atmosphere**.

大气层中有氧气，氧气是人类和其他动物呼吸所必需的气体。

These gases include **oxygen**, which is the gas that humans and other animals need to breathe.

和地球一样，金星也有一层厚厚的大气层。

Like Earth, Venus also has a thick atmosphere.

不过，在金星上你并不能呼吸，因为它的大气层是有剧毒的。

You couldn't breathe on Venus, though, because the planet's atmosphere is toxic.

此外，金星的大气层还有别的方式置人于死地。

Venus's atmosphere is deadly in another way, too.

它的气压太高了，一秒之内就能将人碾碎！

It's so heavy, it would crush you in less than a second!

金星是太阳系中最热的行星。它白天黑夜都很热，以至一些金属都能在其表面熔化！

Venus is the hottest planet in the solar system. Both day and night are so hot, that some metals would melt on the planet's surface!

一旦太阳的热量到达金星表面，就会被锁在那里！因为金星厚厚的大气层会让热量无法对外散发。

Once heat from the Sun reaches the surface of Venus, it stays there! That's because the thick atmosphere keeps it trapped on the planet.

太阳 The Sun

云层 Clouds

大气层 Atmosphere

金星 Venus

云层下有什么?
What's Beneath the Clouds?

只用望远镜是不可能穿透厚厚的云层和大气层观察到金星的。

It's not possible to see through Venus's thick clouds and atmosphere using a telescope.

所以科学家就使用了雷达设备,他们往这颗行星发射无线电波。

So scientists use **radar equipment** that sends **radio waves** to the planet.

无线电波会在行星表面反射,将信息传回地球上的电脑。

The radio waves bounce off the planet's surface and carry information back to computers on Earth.

通过使用雷达,科学家们发现了金星上有一座山,高达11千米。

Using radar, scientists found out that Venus has a mountain that is 11 kilometers high.

这座山比地球上最高的山——喜马拉雅山还要高!

That's taller than Mount Everest, the tallest mountain on Earth!

它被命名为麦克斯韦山脉。

The mountain is named Maxwell Montes (MAX-well MON-tez).

科学家们用雷达采集的信息绘制了一系列金星表面的图片。

Scientists used information collected by radar to create these pictures of Venus's surface.

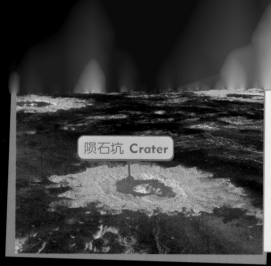

陨石坑 **Crater**

这张图片显示了金星上的一个陨石坑。小行星等大型天体撞击金星表面会留下陨石坑，这些陨石坑的直径能达到好几千米。

This picture shows a **crater** on Venus. Large space objects, such as asteroids, hit Venus's surface and make craters that are many kilometers wide.

玛阿特山 **Maat Mons**

这张图片显示了金星上最高的火山。这座火山有8千米高，名为玛阿特山。

This picture shows Venus's tallest volcano. It is 8 kilometers high. The volcano is called Maat Mons (MAHT MONZ).

探测金星的任务
Missions to Venus

1962年，一个名为"水手2号"的太空飞行器飞越了金星，并将探测信息传回了地球。

这是人类史上第一个成功飞到其他行星的太空飞行器。

自此，有超过20个太空飞行器飞到金星开展研究工作。

其中有些太空飞行器甚至降落到过金星表面。

但是它们很快就被金星的酷热所摧毁，或者被沉重的大气压碾碎！

In 1962, a spacecraft called *Mariner 2* flew past Venus and beamed information back to Earth.

It was the first spacecraft to ever travel successfully to another planet!

Since then, more than 20 spacecraft have flown to Venus to study the planet.

Some of them actually landed on the planet's surface.

The spacecraft were soon destroyed by the heat, though, or crushed by the weight of Venus's atmosphere!

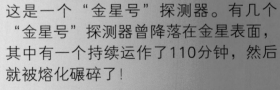

这是一个"金星号"探测器。有几个"金星号"探测器曾降落在金星表面，其中有一个持续运作了110分钟，然后就被熔化碾碎了！

This is a *Venera* spacecraft. Several *Venera* spacecraft landed on Venus. One survived on the planet's surface for 110 minutes. Then it was melted and crushed!

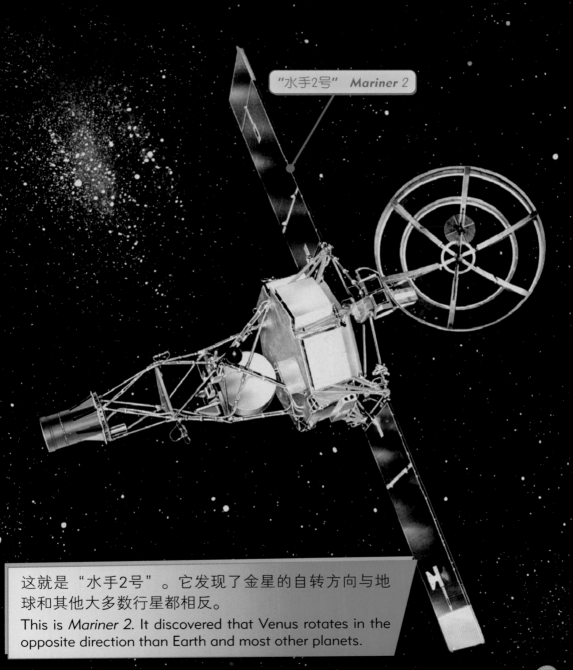

"水手2号" *Mariner 2*

这就是"水手2号"。它发现了金星的自转方向与地球和其他大多数行星都相反。

This is *Mariner 2*. It discovered that Venus rotates in the opposite direction than Earth and most other planets.

探索金星
Exploring Venus

1989年5月，一个名为"麦哲伦号"的太空飞行器飞向了金星。

In May 1989, a spacecraft named *Magellan* flew to Venus.

"麦哲伦号"绕着金星公转，对这颗行星展开了长达4年的研究。

Magellan orbited Venus and studied the planet for four years.

它收集了大量信息，帮助科学家们绘制了有史以来的第一张金星地形图。

It collected information that helped scientists make the first ever map of Venus's surface.

"麦哲伦号"探测任务的最后一部分就是飞向这颗行星的大气层。

The final part of *Magellan's* mission was to fly into the planet's atmosphere.

它直到燃烧殆尽前都还在向地球传输有关金星大气层的各种信息！

The spacecraft sent information about Venus's atmosphere back to Earth until it burned up!

"麦哲伦号" *Magellan*

科学家 Scientist

这张照片显示出科学家们研制"麦哲伦号"的样子。

This photo shows scientists working on *Magellan*.

蓝色区域是低洼地带。
Blue areas are low ground.

绿色区域是中等高度地区。
Green areas are medium-height places.

粉棕色区域是高地，比如山脉。
Pinkish-brown areas are high places, such as mountains.

这张金星地形图是由"麦哲伦号"使用雷达绘制的。这些颜色则是后期通过电脑加上的，以显示地表的不同高度。

This map of Venus's surface was created by *Magellan* using radar. The colors have been added on a computer to show the different heights of the land.

这是"金星快车号"，它曾于2006年至2014年围绕金星公转。它发现金星上曾经有海洋，还发现了一些可能至今仍在喷发的火山。

This is *Venus Express*, which orbited Venus from 2006 to 2014. It discovered that there were once oceans on Venus. It also discovered that volcanoes may be erupting today.

有趣的金星知识
Venus Fact File

以下是一些有趣的金星知识：金星是距离太阳第二近的行星。

Here are some key facts about Venus, the second planet from the Sun.

金星的发现
Discovery of Venus

不用望远镜也能在天空中看见金星。
人们早在古代就发现了金星。

Venus can be seen in the sky without a telescope.
People have known it was there since ancient times.

金星是如何得名的
How Venus got its name

金星是以古罗马爱与美的女神的名字命名的。

The planet is named after the Roman goddess of love and beauty.

行星的大小
Planet sizes

这张图显示了太阳系八大行星的大小对比。

This picture shows the sizes of the solar system's planets compared to each other.

水星 Mercury
地球 Earth
木星 Jupiter
天王星 Uranus
太阳 Sun
金星 Venus
火星 Mars
土星 Saturn
海王星 Neptune

金星的大小
Venus's size

金星的直径约12 104千米

12,104 km across

金星自转一圈需要多长时间
How long it takes for Venus to rotate once

243个地球天

243 Earth days

金星与太阳的距离
Venus's distance from the Sun

金星与太阳的最短距离是107 476 170千米。

金星与太阳的最远距离是108 942 780千米。

The closest Venus gets to the Sun is 107,476,170 km.

The farthest Venus gets from the Sun is 108,942,780 km.

金星围绕太阳公转的平均速度
Average speed at which Venus orbits the Sun

每小时12 6074千米

126,074 km/h

金星绕太阳轨道的长度
Length of Venus's orbit around the Sun

679 892 378千米

679,892,378 km

太阳 Sun

金星 Venus

金星轨道 Venus's orbit

金星上的一年
Length of a year on Venus

225个地球天

225 Earth days

金星的卫星
Venus's moons

金星没有卫星。

Venus has no moons.

金星上的温度
Temperature on Venus

462摄氏度

462°C

动动手吧：制作金星拼贴画
Get Crafty : Make a Hot and Toxic Collage

制作一张拼贴画，展现金星的大气层和云层是如何把太阳牢牢遮挡起来的吧！

天体可以用以下材料来表示：

- 小块的彩色卡纸或纸板
- 餐巾纸或者礼物包装纸

这里有一些金星拼贴画的图片可供参考，当然你也可以创作属于自己的"炎热的剧毒星球"！想想看：

- 你会用什么颜色来展现金星周围的有毒气体呢？
- 你会怎么让金星看起来酷热无比呢？

你需要：

- 一大张薄薄的卡纸或者美术纸（作背景）
- 剪刀
- 白胶水
- 涂胶水用的刷子
- 一位成年人（帮忙裁剪）

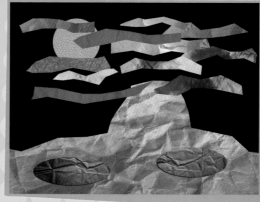

词汇表 Glossary

小行星 ｜ asteroid
围绕太阳公转的大块岩石，有些小得像辆汽车，有些大得像座山。

大气层 ｜ atmosphere
行星、卫星或恒星周围的一层气体。

彗星 ｜ comet
由冰、岩石和尘埃组成的天体，围绕太阳公转。

陨石坑 ｜ crater
圆形坑洞，通常由小行星和其他大型岩石天体撞击行星或卫星表面而形成。

矮行星 ｜ dwarf planet
围绕太阳运行的圆形或近圆形天体，比八大行星小得多。

气体 ｜ gas
无固定形状或大小的物质，如氧气或氦气。

公转 ｜ orbit
围绕另一个天体运行。

氧气 ｜ oxygen
空气中一种无形的气体，是人类和其他生物呼吸所必需的。

行星 | planet

围绕太阳公转的大型天体：一些行星，如地球，主要是由岩石组成的；其他的行星，如木星，主要是由气体和液体组成的。

雷达设备 | radar equipment

能够发射无线电波的机器。当无线电波从物体上反弹回来的时候，人们就用它来分析信息，确定该物体的体积、形状等。

无线电波 | radio wave

一种不可见的波，能够穿过空气执行各种任务。比如说，你的遥控器就能向电视发送无线电波，从而转换频道。

自转 | rotate

物体自行旋转的运动。

太阳系 | solar system

太阳和围绕太阳公转的所有天体，如行星及其卫星、小行星和彗星。

剧毒的 | toxic

毒性很强的，会对生物体造成严重伤害的。

火山 | volcano

地下岩浆喷出地表形成的山丘，部分火山会有高温的液态岩石和气体从开口处喷发；存在于行星或其他天体上。